The Power of Hy

Unlocking the Secrets of the Subconscious Mind

Richard J. Kaspar

Title: The Power of Hypnosis

Author: Richard J. Kaspar

Copyright © Richard J. Kaspar

All rights are reserved by the author. No part of this book may be reproduced without the permission of the Author.

First edition: January 2024

Disclaimer

The information in this book is provided for informational and educational purposes and is not a substitute for professional medical advice, diagnosis, or treatment. If you have concerns or questions about your health, it is essential that you consult with a physician or other qualified health care provider.

Although every effort has been made to ensure the accuracy of the information, the author and publisher disclaim any liability for errors, omissions, or adverse effects resulting from the use of the information contained in this book.

Index

- The magic of hypnosis: a brief history.
- The purpose of this book.

1. ***What is hypnosis?***
- Definition and common misconceptions.
- The historical evolution of hypnosis.
- Difference between hypnosis and sleep.

2. ***The brain and hypnosis.***
- Brain structure and relevant functions.
- States of consciousness: from the alert state to the hypnotic state.
- Altered states of consciousness.

3. ***The power of the subconscious mind.***
- What is the subconscious mind?
- How hypnosis can access the subconscious mind.
- The subconscious mind as a resource.

4. *Hypnotic induction techniques.*
- Progressive induction.
- Rapid induction.
- Instantaneous induction.

5. *Therapeutic applications of hypnosis.*
- Hypnotherapy: definition and scope.
- Case histories: anxiety, phobias, addictions, and more.
- Real case study.

6. *The limits and dangers of hypnosis.*
- Myths and realities.
- Hypnotizing against a person's will.
- Side effects and precautions.

7. *Conclusion*
- The future of hypnosis and subconscious discovery.
- Invitation to responsible and conscious practice.

Introduction

The magic of hypnosis: a brief history.

Hypnosis has always held a mysterious fascination for the collective imagination. Images of eyes swinging back and forth, pendulums swinging and people carrying out commands seemingly against their will have populated pop culture for decades. But where did this powerful practice originate and how has it evolved over the centuries?

The roots of hypnosis can be traced back to ancient civilizations. The ancient Egyptians, for example, had sleep temples, where priests performed rituals that induced trance states in individuals seeking healing or visions. These sacred rituals were based on the belief that entering an altered state of consciousness could allow access to healing powers or divine knowledge.

Even in ancient Greece, sacred places such as the Oracle of Delphi were the scene of practices that can be compared to hypnosis. Pythia, the priestess of the oracle, would enter deep trances, during which she would utter prophecies.

With the rise of modern science, hypnosis began to gain recognition as a legitimate practice with potential therapeutic benefits. In the 18th century, Franz Mesmer, an Austrian physician, proposed the theory of "animal magnetism," arguing that there was an invisible force or fluid within living organisms that could be manipulated for healing. Although Mesmer's theories were later discredited, he set the stage for further research on hypnosis and its effects on the body and mind.

The 19th century saw a growing interest in hypnosis, with pioneers such as James Braid, who coined the term "hypnosis" inspired by the Greek word "hypnos," meaning sleep. Braid also rejected Mesmer's idea of animal magnetism, proposing instead that hypnosis was a state of concentration or focused attention.

Throughout the 20th century, hypnosis found application in various fields, from medicine to psychology. Both Freud and Jung recognized the value of hypnosis, although they had different views on its use in therapy.

Today, hypnosis is recognized as a valuable tool in many therapeutic disciplines, used both to help people overcome trauma and phobias and to enhance motivation, reduce stress, and many other benefits.

In conclusion, the history of hypnosis is fascinating and complex, an interweaving of myths, sacred rituals, science and therapy. Although it has often been shrouded in mystery and misunderstanding, it has stood the test of time, testifying to its power and ability to unlock the secrets of the human mind.

The purpose of this book.

In a world where we are constantly bombarded with information, stimuli and distractions, it becomes increasingly difficult to connect with the deepest and most hidden part of ourselves: our subconscious mind. Hypnosis, through the ages, has emerged as a powerful tool for accessing this inner realm, allowing us to understand, heal and empower ourselves in ways that might otherwise remain unexplored.

The purpose of this book is not only to provide a historical and scientific overview of hypnosis, but also to reveal how this practice, when used correctly, can transform our lives. Each chapter is written with the intention of guiding the reader on a journey of discovery: from understanding the brain mechanisms underlying hypnosis to the actual practice and its therapeutic applications.

We want to:

- *Demystifying Hypnosis*: Removing misconceptions and myths surrounding hypnosis by providing a clear understanding of what it really is and how it works.
- *Exploring the Subconscious Mind*: Shedding light on the powerful resource that is our subconscious mind and how hypnosis can serve as a key to access it.
- *Providing Practical Tools*: In addition to theory, this book aims to provide concrete techniques and strategies for those who wish to explore hypnosis, both as an independent practice and as a therapy.
- *Promote the ethical use of hypnosis*: Emphasize the importance of a responsible and conscious approach to hypnosis, always keeping the well-being of the subject in mind.

In conclusion, this book is intended for anyone who is curious to learn more about hypnosis, whether you are a neophyte looking for an initial introduction or a professional who wishes to deepen your knowledge.

Through the pages that follow, we hope to offer a comprehensive, enlightening and practical overview of the transformative power of hypnosis and the magic hidden in our subconscious mind.

CHAPTER 1

What is hypnosis?

Definition and common misconceptions.

Hypnosis, in all its nuances and complexities, has often aroused curiosity, interest and, at times, suspicion. Before we delve into its many applications and benefits, it is essential to have a clear understanding of what it really is and what misconceptions surround it.

- ***Definition of hypnosis.***

Hypnosis can be defined as an altered state of consciousness in which the subject experiences intense focus, deep relaxation and increased receptivity to suggestions. It is a state in which the conscious mind (the logical and analytical part) takes a step back, allowing the subconscious mind to emerge and become more influential.

- ***Common misconceptions.***
  - ***Hypnosis as mind control***: One of the most persistent myths about hypnosis is that it can be used to control people's minds, making them slaves to the will of the hypnotist. In reality, during hypnosis, the subject always remains in

control and cannot be induced to do anything that goes against his or her fundamental principles or desires.

- *You can get "trapped" in hypnosis*: Some people fear that once they enter a hypnotic state, they may never wake up again. This fear is unfounded. Even if the hypnotist were to leave, the subject would simply awaken naturally or pass into a state of normal sleep.

- *Hypnosis as deep sleep*: Although the root of the word "hypnosis" is derived from the Greek word "hypnos," meaning sleep, hypnosis is not sleep in the traditional sense. Rather, it is a state of intense focus, although it may appear as sleep because of the deep relaxation that often accompanies trance.

- *Only "weak-minded" people can be hypnotized*: A mistaken belief is that only those who are easily influenced or "weak-minded" can be hypnotized. In reality, the ability to enter hypnosis has nothing to do with mental weakness. On the contrary, people with strong concentration and imagination often find hypnosis particularly effective.

Incorporating an accurate understanding of hypnosis and dispelling these common myths is essential to fully appreciate the therapeutic and transformative potential of this practice. With a clear understanding of what hypnosis is and what it is not, we can proceed with an open mind ready to explore the deep recesses of our subconscious.

The historical evolution of hypnosis.

Hypnosis, although often associated with contemporary practices of psychotherapy or entertainment, has its roots in ancient times, crossing cultures and civilizations, manifesting itself in various ways and forms.

- **First tracks.**

    The earliest evidence of hypnosis-like practices can be traced back to ancient civilizations. Shamans from different cultures would enter trance states during rituals, acting as a bridge between the physical and spiritual worlds. These altered states of consciousness were considered sacred and were used for healing, divination or connection with spirits.

- **Egypt and ancient Greece.**

    As mentioned earlier, ancient Egypt had its "temples of sleep," where priests used induction techniques to bring individuals into a trance state. Ancient Greece also possessed its shrines, such as the Oracle of Delphi, where priestesses entered deep trance states to receive messages from the gods.

- ***The Middle Ages and the Dawn of Science.***

  During the Middle Ages, practices that evoked altered states of consciousness were often viewed with suspicion and associated with witchcraft or heresy. However, with the Renaissance and the advent of scientific thought, there was a renewed interest in the human mind and its potential.

- ***Franz Mesmer and "Animal Magnetism.***

  The 18th century saw the figure of Franz Mesmer, who, with his theory of "animal magnetism," gave a kind of modern revival to hypnosis. Although his theories were later questioned, Mesmer paved the way for a more scientific understanding of hypnosis.

- ***The golden age of hypnosis.***

  The 19th and early 20th centuries can be considered the golden age of hypnosis. With figures such as James Braid, Jean-Martin Charcot, and Sigmund Freud, hypnosis began to gain a place in medicine and psychology. Braid, in particular, helped detach hypnosis from Mesmer's theories and establish more scientific foundations.

- ***Modern hypnosis.***

  Throughout the 20th century and to the present day, hypnosis has found its way into various areas, from therapy to pain management, from sports training to self-help. Research continues to evolve, and with it our understanding of hypnosis and its infinite applications.

The history of hypnosis is as rich as it is fascinating. From ancient sacred rituals to modern therapeutic tool, hypnosis has spanned centuries, withstanding challenges, misconceptions and cultural changes. With this deep historical understanding, we can truly appreciate its importance and place in the landscape of treatment and understanding of the human mind.

Difference between hypnosis and sleep.

For many, the word "hypnosis" conjures up images of deeply asleep individuals, seemingly suspended between reality and another world. However, it is critical to understand that hypnosis and sleep are two distinct states, although they may share some superficial similarities. Let us explore the key differences between these two states.

- ***Origin of the term.***

    The confusion between hypnosis and sleep may have its etymological roots. The term "hypnosis" is derived from the Greek word "hypnos," meaning "sleep." This nomenclator was introduced by James Braid, one of the pioneers of modern hypnosis, who later attempted to change the term when he better understood the nature of the phenomenon. However, the name "hypnosis" remained.

- ***Characteristics of the hypnotic state.***
    - ***During hypnosis***: The person remains conscious and aware of his or her

environment, although he or she may feel deeply relaxed.

There is intense internal focus and increased susceptibility to suggestions.
- Cognitive processes such as memory and imagination can be amplified.
- The person can communicate, answer questions and follow instructions.
- Sleep state characteristics.

- ***During sleep***:
- The person is generally unaware of his or her surroundings.
- Cognitive processes are reduced and the mind becomes immersed in dreams.
- Conscious communication is absent; a sleeping person cannot consistently answer questions or follow instructions.

- Sleep has specific stages, including REM sleep and non-REM sleep, each with its own physiological and neurological characteristics.

- ***Why is the distinction important?***

Knowing the difference between hypnosis and sleep is essential for many reasons:

- ***Therapy:*** In hypnotherapy, the therapist needs the subject to be in a state of hypnosis, not sleep, in order to work effectively with the subconscious mind.
- ***Safety:*** A person in hypnosis can be "awakened" very easily and can react quickly to emergencies, unlike someone who is sound asleep.
- ***Research:*** In hypnosis research, it is vital to differentiate between data obtained from subjects in the hypnotic state and those in the sleep state, as the two states have different impacts on brain patterns and physiology.

Hypnosis and sleep may seem similar at first glance, but it is essential to recognize them as separate entities to fully appreciate the potential and limitations of each.

CHAPTER 2

The brain and hypnosis.

Brain structure and relevant functions.

To fully understand hypnosis, we must first explore the complex structure of our brain and the relevant functions it performs. Not only is the brain the center of our thinking, but it is also the hub of emotions, perceptions, and responses to the external world. Here is an overview of the crucial areas of the brain and their relationship to hypnosis.

- *Cerebral cortex.*

The cerebral cortex is the outer part of the brain and is responsible for high-level thinking functions, such as reasoning, perception, and planning. It is divided into lobes:

- *Frontal lobe*: Responsible for planning, reasoning and impulse control.
- *Temporal lobe*: Manages hearing and long-term memory.
- *Parietal lobe*: Deals with spatial perception and cognition.
- *Occipital lobe*: Centered on vision.

During hypnosis, the cortex may show changes in activity, especially in areas related to attention and focus.

- ***Limbic system.***

Located deeper in the brain, the limbic system manages emotions and memory. It includes:

- *Amygdala*: Regulates emotional responses, particularly fear.
- *Hippocampus*: Central to memory and learning.

Hypnosis can affect the limbic system, especially when used to deal with trauma or painful memories.

- ***Thalamus.***

The thalamus acts as a "switchboard" for sensory signals, sending information to the appropriate parts of the cerebral cortex. During hypnosis, the thalamus can help filter out distractions, allowing the subject to better focus on the hypnotist's suggestions.

- ***Cerebellum.***

Located at the base of the brain, the cerebellum coordinates movement and balance. Although not directly involved in hypnosis, it plays a role in regulating posture and muscle relaxation.

- *Activating reticular system (SRA).*

SRA regulates wakefulness and sleep, and is essential for maintaining the state of consciousness. During hypnosis, SRA helps to keep the subject in a state of vigilant relaxation.

The brain is a complex and dynamic organ, and hypnosis interacts with several of its parts in unique ways. Understanding the structure and functions of the brain gives us a clearer view of how hypnosis can affect our minds and bodies, providing a solid foundation for exploring the therapeutic and practical applications of this fascinating technique.

States of consciousness: from the awake state to the hypnotic state.

States of consciousness represent the different ways through which we perceive, interact and react to the world around us. From active daytime awareness to nighttime dreams, our consciousness fluidly shifts between different phases. But where does hypnosis fit into this continuum? Let's explore the various states of consciousness and see how hypnosis fits into the big picture.

- ***Vigilant State.***

This is the state of consciousness we experience in daily life while awake and active. In this state:

- The perception of the outside world is acute.
- Rationality and critical thinking are at an all-time high.
- We are able to respond quickly to stimuli.

- ***State of Relaxation.***

As we relax, perhaps meditating or listening to quiet music:

- It decreases our awareness of external stimuli.
- More introspection occurs.
- Thoughts may become more floating or dreamy.
- **Light Sleep.**

At this preliminary stage of sleep:

- Awareness of the outside world decreases further.
- The body begins to relax deeply.
- Thoughts become even more blurred.
- **Deep Sleep.**

As we plunge into deep sleep:

- Consciousness of the outside world is almost totally absent.
- Physical and mental regeneration is at its peak.
- You may experience vivid dreams, especially during REM phase.
- **Hypnotic State.**

Hypnosis sits at a unique point between these states:

- Awareness of the outside world is reduced, but not completely absent.
- There is a deep relaxation and openness to suggestions.
- The mind is focused and highly susceptible, allowing direct access to subconscious thoughts.

Our mind is a dynamic entity, capable of moving through various states of consciousness in response to internal and external stimuli. Hypnosis occupies a unique place in this continuum, offering a bridge between active awareness and the deeper recesses of the mind. Understanding how hypnosis relates to other states of consciousness helps us not only to demystify this practice, but also to better understand the incredible adaptability and depth of the human mind.

Altered states of consciousness.

Altered states of consciousness represent modes of perception and interaction that deviate from the normal waking experience. These states can result from natural causes, meditative practices, traumatic experiences or the use of psychoactive substances. But what exactly does "altered" mean? And how does hypnosis fit into this picture? Let's explore the concept of altered consciousness and its various aspects.

- ***Definition of Altered States of Consciousness.***

An altered state of consciousness occurs when:

- There is a significant change in the perception of time, space and self.
- There are variations in the quality or clarity of thought.
- Unusual sensations, images or sounds are experienced.

- ***Meditation and Trance.***

Meditation, practiced in many cultures and spiritual traditions, can lead to deep states of relaxation and altered awareness. Trance, often associated with religious rituals or dances, represents another state of intense focus and altered perception.

- ***Dreams and Lucid Dreaming.***

Dreams are natural altered states that we experience every night. "lucid dreaming," or lucid dreaming, represents an interesting variation in which the dreamer is aware that he or she is dreaming and can exercise some degree of control over the dream itself.

- ***Substance-Induced Experiences.***

The use of psychoactive substances, such as hallucinogens, cannabis or MDMA, can induce marked altered states characterized by altered visual or auditory perceptions, feelings of universal connectedness or deep introspection.

- ***Trauma and Dissociation.***

Traumatic events can cause altered states of consciousness in the form of dissociation, where the individual feels detached from the surrounding reality or his or her body. This may be a defense mechanism to protect oneself from emotionally painful experiences.

- ***Hypnosis as an Altered State.***

Hypnosis shares some characteristics with altered states of consciousness:

- A change in the perception of time and space.
- Intense internal focus.
- An increased susceptibility to suggestions.

However, hypnosis is unique in that it is inducible, controllable, and can be used therapeutically.

Altered states of consciousness represent a vast and intriguing field of study, allowing us to explore the flexibility and depth of the human mind. Whether in a drum-induced trance, lucid dreaming or hypnotic experience, these states offer us a window into aspects of consciousness beyond our everyday waking experience, expanding our understanding of human nature.

CHAPTER 3

The power of the subconscious mind.

What is the subconscious mind?

Many times in daily life, we find ourselves performing actions or experiencing emotions without a full awareness of what drives us. These hidden forces that guide our behaviors and reactions are rooted in the subconscious mind. But what exactly does this level of consciousness represent? And why is it so influential in our lives?

- ***Definition of the Subconscious Mind.***
  The subconscious mind represents that part of our consciousness that operates below the level of immediate awareness. It is the repository of our memories, emotions, beliefs and habits, many of which were formed during our childhood.
- ***The Role of Memories.***
  From first love to traumatic experiences, our subconscious mind holds memories that we may have forgotten on a conscious level, but which continue to influence our actions and reactions.
- ***Emotions and Impulses.***

The subconscious mind is also the seat of our emotions. Although we may not always be aware of what we feel, hidden emotions can profoundly influence our moods and behaviors.

- ***Convictions and Mental Schema.***

  From birth, we begin to form beliefs about the world and ourselves. Many of these beliefs, which can be both limiting and empowering, reside in the subconscious mind and influence our self-perception and the decisions we make.

- ***Habits and Automatic Behaviors.***

  Many of the behaviors we repeat daily, such as driving or making coffee, are governed by the subconscious mind. These actions become so automatic that we often do not realize we are performing them.

- ***The Relationship with the Conscious Mind.***

  While the conscious mind is the locus of reflection, critical thinking and decision making, it operates in tandem with the subconscious mind. Many of the "intuitions" or "instincts" we experience actually come from the subconscious mind.

The subconscious mind is a powerful driving force behind our personality and behavior. It holds a vast network of information, emotions and habits that, although hidden, deeply shape our life experience. Understanding its functioning and influence is critical for those wishing to access tools such as hypnosis in order to catalyze positive change and achieve greater self-understanding.

How hypnosis can access the subconscious mind.

Hypnosis, from its earliest days, has been seen as a mysterious key that can unlock the hidden recesses of the mind. While science has provided illuminations on how hypnosis works, its ability to access the subconscious mind remains one of its most fascinating and powerful aspects. But how does this connection take place? And how can it be used to bring benefit?

- ***Bypassing the Conscious Critic.***

One of the primary goals of hypnotic induction is to "bypass" the conscious critic, that rational, analytical filter that judges, evaluates and often blocks information. Once this filter is reduced, the subconscious mind becomes more accessible and receptive.

- ***Augmented Susceptibility.***

During hypnosis, susceptibility to suggestions is greatly increased. This means that the therapist can communicate directly with the subject's subconscious mind, offering ideas and suggestions that can have a lasting effect.

- ***The Importance of Relaxation.***

A deep feeling of relaxation is often the gateway to the subconscious mind. In the relaxed state, defenses are lowered and the individual becomes more open and receptive, facilitating access to subconscious content.

- ***Visualizations and Symbols.***

The subconscious mind communicates differently from the conscious mind. While the latter uses logic and language, the subconscious mind is rich in symbols, images and emotions. Through guided visualizations, hypnosis can interact with these symbols, offering new perspectives and solutions.

- ***Regression and Hidden Memories.***

One of the most powerful techniques in hypnosis is regression, which allows one to relive and reprocess past events stored in the subconscious mind. This can help understand and resolve trauma or painful experiences that affect the present.

- ***Strengthening New Behavioral Models.***

Hypnosis can not only explore the subconscious mind but also change it. With positive suggestions and reinforcement, new patterns of thought and behavior can be created that are "imprinted" on the subconscious mind, facilitating the desired change.

Hypnosis represents an extraordinary tool for connecting with the subconscious mind, offering opportunities for healing, growth and transformation. Through understanding and using this practice, we can not only access the hidden treasures of our mind, but also free ourselves from old chains and embrace our full potential.

The subconscious mind as a resource

If the subconscious mind were an ocean, most of us would only swim on the surface, never exploring the unknown depths below. But these depths hold priceless treasures: insights, wisdom, hidden talents and more. The subconscious mind is not just a repository of old memories and traumas; it is also an incredibly powerful resource that can enrich and transform our lives.

- *Intuition and Inspiration.*

    Intuition is that visceral feeling or sudden "knowing" that often comes out of nowhere. It comes from the subconscious mind, a place where our brain processes information without our conscious intervention. That's why solutions to problems often "come" when we least expect them.

- *Creativity and Flow.*

    Artists, musicians and creators know well how crucial the subconscious mind is in the creative process. It is the place from which innovative ideas and unique connections emerge. When we are immersed in creativity, we can access a state of "flow," where time seems to stop and we are completely absorbed in the moment.

- ***Healing and Growth.***

The subconscious mind stores not only trauma but also resources for healing. Through therapy, hypnosis and other practices, we can access these resources, address old pains and embark on a path of growth and transformation.

- ***Hidden Skills and Talents.***

Many of us have untapped talents and abilities that lie dormant in the subconscious mind. Whether it is an artistic talent, a language skill or a forgotten passion, the subconscious mind can help us rediscover and cultivate these hidden gems.

- ***Autosuggestion and Self-Realization.***

The subconscious mind is incredibly sensitive to suggestions. This can be used to our advantage through autosuggestion techniques, helping us to strengthen self-confidence, overcome fears and achieve the goals we desire.

- ***Connection with the Collective.***
Some theorize that the subconscious is also a bridge to the collective consciousness, a kind of "network" that connects all human beings. If this were true, we could tap into a wisdom and connection far greater than ourselves.

The subconscious mind is not just a place of shadow and mystery, but an inexhaustible source of potential and possibility. By treating it as a valuable resource and learning how to work with it, we can unlock new dimensions of ourselves and live richer, more satisfying lives aligned with our true purpose.

CHAPTER 4

Hypnotic induction techniques.

Progressive induction.

Hypnosis, like any art, has a variety of techniques and methods to achieve its goal. One of the best known and most widely used techniques is progressive induction. This method, based on gradual relaxation of the body and mind, is essential for anyone who wishes to understand or practice hypnosis.

- ***What is Progressive Induction?***

    Progressive induction is a methodical and gradual approach to bringing a person into a hypnotic trance state. This technique involves a series of suggestions and instructions aimed at progressively relaxing each part of the body and, subsequently, the mind.

- ***The importance of Relaxation.***

    Relaxation is the cornerstone of progressive induction. Before accessing the subconscious, it is essential that the subject feel completely at ease and free of any tension. Physical relaxation often leads to mental relaxation, setting the stage for a deep trance.

- ***Step by Step: The Sequence.***
  Progressive induction usually takes place following a precise sequence, starting at the tip of the toes and moving up to the head. As each part of the body is relaxed, the subject becomes increasingly receptive to subsequent suggestions.
- ***The Voice and the Guide.***
  The voice of the therapist or hypnotist plays a crucial role in this process. A calm, steady, reassuring tone helps the subject trust and follow instructions. The voice becomes a guide, leading the subject through the depths of relaxation.
- ***Deepening Trance.***
  Once complete relaxation has been achieved, the hypnotist can use various methods to further deepen the trance state, such as imagining a ladder or elevator descending deeper and deeper.
- ***Benefits of Progressive Induction.***
  In addition to being an effective technique for achieving hypnotic trance, progressive induction can have numerous other benefits, such as

reducing stress, improving sleep, and enhancing body awareness.

Progressive induction represents one of the fundamental techniques in a hypnotist's arsenal. Its effectiveness lies in its simplicity and its ability to bring the subject into a state of deep relaxation and receptivity. Whether you are a professional or newbie to hypnosis, understanding and mastering this technique can open the door to profound and lasting transformations.

Rapid induction.

While progressive induction is a slow, meticulous dance that guides the subject into a trance state, rapid induction is a daring leap into the heart of trance. This technique, often associated with theatrical demonstrations and show magicians, is powerful and immediate, but requires special mastery and insight.

- **What is Rapid Induction?**

  Rapid induction is a technique to bring a person into a deep trance state in a very short time, often in less than a minute. Instead of gradually relaxing the subject, rapid induction uses direct suggestions and sometimes physical or mental shocks to quickly induce trance.

- **Shock as a gateway.**

  A key component of rapid induction is the element of surprise or shock. This can be achieved by an interrupted handshake, a sudden jerk, or a firm verbal command. Shock serves to temporarily disorient the conscious mind, creating a brief window to insert a hypnotic suggestion.

- *Trust and Connection.*

   For rapid induction to be successful, it is essential to establish a bond of trust with the subject. If the subject trusts the hypnotist, he or she will be more receptive to quick and direct suggestions, making the induction more effective.

- *Advantages of Rapid Induction.*

   The main advantage of rapid induction is, of course, its speed. This makes it particularly useful in situations where time is of the essence, such as in live demonstrations or in some therapeutic settings.

- *Therapeutic uses*

   Despite its association with stage hypnosis, rapid induction also has therapeutic applications. In the hands of an experienced therapist, it can be an effective way to help clients overcome trauma, fears, or blocks quickly and directly.

- *Limitations and precautions*

   Rapid induction is not suitable for everyone. Some people may find it too intrusive or disconcerting. Also, without proper training and understanding, there is a risk of provoking anxiety or resistance in the subject.

Rapid induction is a fascinating technique that demonstrates the amazing malleability of the human mind. While it may seem magical or mysterious, it is based on sound psychological principles. Like all hypnotic techniques, it requires respect, understanding and practice to be used effectively and responsibly.

Instant Induction.

Instantaneous induction represents one of the pinnacles of talent and dexterity in the art of hypnosis. While progressive and rapid induction techniques rely on relaxation or the element of surprise, instantaneous induction, as the name suggests, brings the subject into a trance state in the blink of an eye. But how is this possible, and what are the implications of such an ability?

- **What is Instant Induction?**

    Instantaneous induction is a technique that allows the hypnotist to guide the subject into a deep trance state almost immediately, often within seconds. This technique combines elements of shock and suggestion to create an instantaneous hypnotic response.

- **The Art of Suggestion.**

    Central to instant induction is the power of suggestion. The hypnotist gives a command or series of commands with a certainty and authority that leaves no room for doubt in the subject's mind.

- ***Mechanics of Induction.***

    Similar to rapid induction, shock can also be used here, but the speed and effectiveness of the suggestion are crucial. The hypnotist might use a key word, a gesture or a combination of both to achieve the desired effect.

- ***Safety and Accountability.***

    Because of the rapidity and intensity of instantaneous induction, it is critically important that it be practiced in a safe environment and with great responsibility. Not everyone will react the same way, and a mishandled induction could cause stress or anxiety in the subject.

- ***Practical Uses of Instant Induction.***

    Although it may seem like a technique suitable only for stage hypnosis, instant induction has practical applications in therapy as well. It can be used to help patients quickly confront fears or trauma, or to institute positive conditioned responses.

- ***The importance of Practice.***
  Mastery in instant induction requires practice and experience. The ability to read and interpret the subject's responses, to adapt to different situations, and to lead with confidence and authority are all qualities that are refined with time.

Instant induction unveils the true potential of the human mind and hypnosis' ability to quickly achieve deep states of awareness. While it may seem amazing and spectacular, like all techniques, it requires respect, dedication and an ethical approach to ensure the well-being and safety of the subject.

CHAPTER 5

Therapeutic applications of hypnosis.

Hypnotherapy: definition and scope.

If hypnosis is the art of guiding the mind into an altered state of consciousness, hypnotherapy is the art of using that state to promote healing, change and growth. This subchapter is devoted to exploring the definition of hypnotherapy and its broad fields of application.

- **What is Hypnotherapy?**

  Hypnotherapy is a form of therapy that uses hypnosis as the main therapeutic tool. Through hypnotic trance, the therapist can access the depths of the patient's subconscious, where ingrained beliefs, emotions, and behavioral patterns reside.

- **History of Hypnotherapy.**

  While hypnosis as a practice has ancient roots, hypnotherapy as a structured profession began to gain popularity in the 20th century. Physicians and psychologists recognized the potential of hypnosis as a therapeutic tool and began to develop specific techniques and protocols.

- **Advantages of Hypnotherapy.**

One of the main advantages of hypnotherapy is its ability to bypass cognitive resistance and directly access the subconscious mind. This allows it to work on problems and traumas that other therapies might find difficult to address.

- *Scope.*

Hypnotherapy has a wide range of applications, including:

- *Pain management*: Reduction of chronic or acute pain.
- *Anxiety disorders*: Treatment of phobias, panic attacks, and post-traumatic stress.
- *Addictions*: Helping people quit smoking or manage other forms of addiction.
- *Performance Enhancement*: Helping athletes, artists and professionals improve their performance.
- *Emotional issues*: Working on trauma, grief, low self-esteem and other emotional issues.

- *The difference with other Therapies.*

While many therapies focus on conversation and conscious awareness, hypnotherapy is distinguished by its direct access to the subconscious mind. This unique approach allows for faster and deeper results in many circumstances.

Hypnotherapy represents a bridge between ancient trance rituals and modern mind sciences. With a growing base of scientific research supporting its effective techniques, hypnotherapy continues to offer a valuable resource for those seeking change, healing and inner understanding.

Case histories: anxiety, phobias, addictions, and more.

Hypnotherapy is not just an abstract theory or esoteric practice; it is a powerful technique that has transformed the lives of many people. In this sub-chapter, we will explore specific case histories in which hypnotherapy has demonstrated its effectiveness, offering relief and healing in various areas.

- *Anxiety:* Living with a Sense of Constant Worry.
  Marta, a 28-year-old young woman, was constantly burdened by a shadow of anxiety that made her fear the worst in every situation. With hypnotherapy, she discovered that her anxiety was related to childhood trauma. By working on subconscious memories, she was able to scale back her worry and live with greater peace of mind.

- *Phobias:* The Terror of Flying.
  David avoided airplanes like the plague. The mere thought of flying caused him palpitations.

Through hypnotherapy, he revealed a submerged memory of a trauma related to childhood in which he had been trapped in an enclosed place. By connecting and resolving this trauma, David was able to take flight for the first time in years without fear.

- *Addictions:* The Endless Cycle of Cigarettes.

  Lorena had tried everything to quit smoking: patches, gum, even traditional therapies. But with hypnotherapy, she identified the link between smoking and the feeling of comfort she sought because of unresolved family problems. Once this root was addressed, quitting became much easier.

- *Trauma:* The Past that Doesn't Pass.

  Roberto, a war veteran, was plagued by recurring nightmares. Through hypnotherapy sessions, he dealt with his wartime traumas, finding a way to process them and free himself from the painful memories.

- ***Self-Esteem and Personal Growth.***
  Elisa felt that she could never measure up. Regardless of her achievements, she always saw herself as insufficient. Hypnotherapy revealed that her beliefs stemmed from an inferiority complex rooted in childhood. By acknowledging and reframing these beliefs, Elisa was able to embrace her true worth.

These are just a few examples that show the wide range of applications of hypnotherapy. Each individual carries a unique mosaic of experiences, traumas and beliefs. Hypnotherapy offers a valuable tool to decipher this mosaic and help people find peace, understanding and transformation.

Real case study.

Beyond the theories and principles, the power of hypnotherapy lies in the real-life stories of those who have benefited from its techniques. In this sub-chapter, we will examine a number of real-life cases, recounting their challenges, their journey through hypnotherapy, and the results they achieved.

- *Lucia:* The Ghost of the Past.
  Lucia, 35, had suffered from seemingly unexplained anxiety attacks for years. During hypnotherapy sessions, she revisited a traumatic incident from her childhood that she had removed from her conscious memory. Through guided sessions, Lucia was able to process and release the trauma, allowing her to live a life free of crippling anxiety.
- *Mark*: The Battle with Addiction.
  Marco was a recovering alcoholic. Although he had stopped drinking, the temptation was always there, lurking. Hypnotherapy revealed that Marco was using alcohol as a means of coping with a

series of childhood traumas. With the support of his therapist, Marco dealt with these traumas, significantly reducing his urge to drink and strengthening his path to recovery.

- *Clare:* Phobia of the Dark.

  Clare, a young mother, had a paralyzing fear of the dark. This phobia had become particularly problematic now that she had a child to care for at night. During hypnotherapy sessions, it became apparent that her phobia was related to a frightening event she experienced when she was six years old. By reframing this event, Clare was able to overcome her phobia and live a normal life.

- *Andrea:* Fighting Obesity.

  Andrea had tried every diet imaginable, but could not maintain a healthy weight. Hypnotherapy revealed that Andrea was overeating as a means of coping with feelings of inadequacy and abandonment dating back to her adolescence. By working through these emotional issues, Andrea developed a healthier relationship with food and began a sustainable weight loss journey.

- ***Sofia***: Trauma from Accident.

    After a serious car accident, Sofia had become extremely anxious about driving. Hypnotherapy helped her revisit the accident, not to relive the trauma, but to process it and focus on her ability to overcome and move forward. Gradually, Sofia regained the confidence to return to driving.

These cases illustrate the versatility and depth of hypnotherapy as a healing tool. Each individual has a unique story, and hypnotherapy offers a means to explore, understand, and resolve the personal challenges each of us may face.

CHAPTER 6

The limits and dangers of hypnosis.

Myths and realities.

Hypnosis, with its mysterious appeal and history full of legends, has generated numerous myths and misconceptions over the years. Many of these myths, unfortunately, have hindered the understanding of hypnosis as a therapeutic tool and generated unwarranted misconceptions. In this subchapter, we will separate fact from fiction, revealing the truth behind the most common myths about hypnosis.

- *Myth*: The hypnotized person loses control over himself.
- *Reality*: One of the most common fears about hypnosis is that the hypnotized person loses control of his or her actions or thoughts. In reality, those under hypnosis always remain aware and maintain control of their actions. Hypnosis is a state of relaxation and focus, not loss of control.

- *Myth:* Only weak-minded people can be hypnotized
- *Reality:* The ability to enter a state of hypnosis has nothing to do with intelligence or willpower. In fact, people with strong concentration skills and vivid imaginations often find it easier to enter a hypnotic state.
- *Myth:* The hypnotist has special powers
- *Reality:* Hypnotists do not have magical or supernatural "powers." They are professionals trained to guide people into a state of deep relaxation and to use suggestion techniques to aid in the desired change.
- **Myth:** Hypnosis is a dangerous practice
- *Reality:* When practiced by qualified professionals, hypnosis is a safe procedure. As with any therapeutic treatment, there are potential risks, but these are minimal when hypnosis is used ethically and competently.
- *Myth*: Once hypnotized, one can get stuck in that state.

- ***Reality***: It is impossible to "get stuck" in a hypnotic state. At most, a person might fall asleep and then wake up normally. If a hypnotist were to interrupt a session without "awakening" the subject, the subject would spontaneously emerge from the hypnotic state after a short time.

Understanding the difference between the myths and reality of hypnosis is crucial to fully appreciate its therapeutic value. By separating fact from fiction, we can approach hypnosis with an open mind, ready to harness its potential for well-being and personal transformation.

Hypnotizing against a person's will.

The idea that someone can be hypnotized against his or her will is deeply rooted in popular culture, fueled by film portrayals, television programs and stories. But how much of this corresponds to reality? In this sub-chapter, we explore this controversial topic to outline the truth behind forced or nonconsensual hypnosis.

- ***Consent in Professional Practice.***
  In the professional therapeutic context, consent is a key element. Ethical and properly trained hypnotists will never attempt to hypnotize someone without their explicit consent. Respect for the patient's autonomy and will is central to the practice of hypnotherapy.

- ***Subconscious Rendition and Suggestion.***
  While it is true that hypnosis exploits suggestion and subconscious receptivity, this does not mean that a person can be hypnotized against his or her will. Even in a state of deep relaxation and openness, a person's subconscious has defense and protection mechanisms that prevent actions or

suggestions that go against one's deepest values or desires.

- ***Shows and Performances.***

  Many of the misconceptions about hypnosis that pervade popular culture come from stage performances or television programs. In these contexts, it might appear that people are being hypnotized against their will. However, it is important to remember that these are entertainments and that participants, although they might be surprised by the depth of their experience, have nonetheless given consent to participate.

- ***Hypnosis and Manipulation.***

  There are stories of individuals using suggestion or manipulation techniques to influence others to their advantage. While these tactics may resemble hypnosis, they often do not involve a true hypnotic state. Rather, they rely on power dynamics, deception, and coercion. It is essential to distinguish between these manipulative behaviors and true therapeutic hypnosis.

- ***The Defense of Personal Autonomy.***
Recognizing the power and responsibility associated with hypnosis also means defending the autonomy and right of each individual to make informed choices about his or her own well-being. Educating the public about the realities of hypnosis and emphasizing the importance of consent is essential to maintaining the integrity of this practice.

Side effects and precautions

Hypnosis, when used correctly and by qualified professionals, is considered a safe and noninvasive practice. However, as with any therapeutic intervention or technique, there are potential side effects and precautions to consider. This subchapter explores these issues, helping readers make informed choices about their own mental health and well-being.

- ***Common side effects***
  After a hypnosis session, some people may experience:
- ***Fatigue or drowsiness***: Hypnosis is a deep state of relaxation, and may leave some people feeling tired or drowsy after the session.
- ***Live memories or emotions***: During hypnosis, dormant memories or emotions may emerge. These may sometimes be intense and unexpected.

- ***Mild headache***: Some individuals may experience mild headaches after a session.

- *Rarer adverse reactions.*

In rare cases, hypnosis may cause:

   - *Anxiety or panic*: If a person has unrecognized fears or anxieties related to the idea of "losing control," he or she may experience anxiety during or after the session.
   - *Dissociative disorders*: Some people may feel detached from reality or have the feeling of "floating" after a session.
- *Precautions.*
   - *Medical* history: It is essential to provide the therapist with a complete medical history, including any mental or neurological disorders. Hypnosis may not be suitable for everyone, particularly those with certain psychiatric conditions.

   - *Qualified therapists*: Ensure that the therapist or hypnotist is properly trained and certified. This

will ensure that sessions are conducted in a safe and professional manner.
- *Safe environment*: Hypnosis sessions should be conducted in a quiet, distraction-free environment where the patient feels safe and relaxed.
- **Contraindications.**

Hypnosis may not be suitable for people with:

- Psychotic disorders, such as schizophrenia.
- Some types of anxiety disorders.
- Substance abuse problems, unless used as part of a larger treatment.

CHAPTER 7

Conclusion

The future of hypnosis and subconscious discovery.

As we close this journey through the world of hypnosis and the deep exploration of the subconscious mind, it is essential to reflect on the future. Where is this extraordinary practice taking us? What does the continued exploration of the depths of our psyche hold for us? In this sub-chapter, we cast our eyes toward the horizon to speculate what the next chapters of this fascinating story will be.

- *New Frontiers in Neurological Research.*

    With the advancement of technology and neuroscience, we are beginning to better understand how hypnosis works at the neurological level. Tools such as functional magnetic resonance imaging (fMRI) and electroencephalography (EEG) are offering unprecedented insight into what happens in the brain during a hypnotic state.

- ***Expansion of Therapeutic Applications.***
  Hypnotherapy has already been shown to be effective in treating a wide range of disorders, from anxiety to addiction. As research progresses, we are likely to discover new ways in which hypnosis can be used to promote well-being and treat various disorders.
- ***A Bridge between Western Science and Eastern Traditions***.
  Meditation and other Eastern practices have many parallels with hypnosis. While both practices seek to achieve an altered state of awareness and access deeper levels of the mind, they come from different traditions and philosophies. The future may see further integration of these two disciplines, creating a bridge between Western science and Eastern spiritual traditions.
- ***Customization and Advanced Techniques.***
  With the advent of artificial intelligence and advanced technology, we may see a new wave of personalized hypnotic techniques. These

techniques would be modeled after an individual's specific needs and reactions, ensuring a deeper and more effective hypnotic experience.

- ***Society and the Normalization of Hypnosis.***
  As understanding and acceptance of hypnosis grows, it is likely that we will see greater integration of this practice into everyday life. This could result in more education about its therapeutic power and greater openness on the part of society as a whole.

The future of hypnosis is bright and promising. As we continue to explore the depths of the subconscious mind and unravel its secrets, we move ever closer to a complete understanding of ourselves and the incredible capabilities of the human mind. This journey toward understanding is endless, and hypnosis remains a valuable tool to guide us along the way.

Invitation to responsible and conscious practice

Over the course of this book, we have navigated through the deep sea of hypnosis, unlocking the secrets of the subconscious mind and discovering its incredible potential. Now, as we approach the end of our journey, it is crucial to reflect on how we apply this knowledge and the responsibilities it entails. This sub-chapter serves as an invitation to approach hypnosis with respect, responsibility and awareness.

- *The Power of Hypnosis.*

    As we have seen, hypnosis is a powerful tool. It can unlock dormant memories, help overcome trauma and create profound changes in an individual's behavior and perception. This power, however, must be carefully managed.

- *The Responsibility of the Therapist and the Practitioner.*

    Any person practicing hypnosis, whether a professional therapist or an enthusiastic beginner, has a responsibility to do so ethically and safely. This means:

- Make sure you have adequate training.
- Fully inform patients or participants what to expect and obtain their consent.
- Avoid inducing false memories or harmful suggestions.

- ***The importance of Patient Autonomy.***
  While hypnosis can guide a person into deep states of relaxation and suggestion, it is essential to respect the patient's autonomy and will. No one should ever feel forced or manipulated into a hypnotic state.

- ***Precautions and Contraindications.***
  As discussed in previous chapters, there are certain situations or conditions for which hypnosis may not be appropriate. It is critical that therapists be aware of and comply with these contraindications.

- ***The Invitation to Awareness.***
  Finally, hypnosis is not only a technique: it is also a journey of inner discovery. As such, it requires

awareness. Awareness of one's intentions, one's actions and the profound impact they can have on others.

Approaching hypnosis with respect, care and awareness not only ensures the safety and well-being of those we work with, but also elevates the practice itself. The human mind is a vast and uncharted territory, and hypnosis offers us a map to navigate it. But as with any map, it is essential to use it wisely. This is our invitation to you: enter the world of hypnosis with your eyes open, your heart attentive and your mind ready to learn and grow.

Milton Keynes UK
Ingram Content Group UK Ltd.
UKHW021051120724
445512UK00017BA/336